Raphaël Ngadi Litadi

Manual for corn growers in the humid tropics

Raphaël Ngadi Litadi

Manual for corn growers in the humid tropics

ScienciaScripts

Imprint

Cover image: www.ingimage.com

This book is a translation from the original published under ISBN 978-3-8416-7234-6.

Publisher:
Sciencia Scripts
is a trademark of
Dodo Books Indian Ocean Ltd. and OmniScriptum S.R.L publishing group

120 High Road, East Finchley, London, N2 9ED, United Kingdom
Str. Armeneasca 28/1, office 1, Chisinau MD-2012, Republic of Moldova, Europe
Printed at: see last page
ISBN: 978-620-8-33727-8

PRODUCER'S MANUAL
OF CORN IN TROPICAL ZONES
WET

Raphaël NGADI LITADI

- E-mail: ngadilitadi@yahoo.fr

Signing sessions

I dedicate this book

To my family, who gave me a dignified education, whose love made me what I am today:

Especially my grandmother, the late Elizabeth PANGA, for the taste for effort she inspired in me, through her rigor. I dedicate this milestone in my life to her memory. I hope that, from the world that is now hers, she will appreciate this humble gesture as proof of gratitude on the part of a grandson who has always prayed for the salvation of her soul.

To my late great-great-uncle Basile NDOUMOU, this is my deep gratitude for your eternal love, may this book be the best gift I can give you.

To all those who have closely or remotely followed the progress of this work and made the necessary corrections at each stage. Please accept my sincere thanks and the expression of my profound gratitude.

FOREWORD

Growing maize in humid equatorial environments has its own particular way of working, linked to very specific climatic conditions.

This manual has therefore been written in response to requests from people interested in this crop, but lacking sufficient knowledge of its technical itinerary.

This technical note on maize production is based on our experience gained at the Société Industrielle d'Agriculture et d'Elevage de Boumango (SIAEB), in village plantations and elsewhere.

Its main aim is to provide those wishing to invest in this crop with the information they need to establish and manage it.

Corn growers will find in this manual useful additions to their information, to simplify their work and improve their yields.

The main difficulty encountered in producing this manual was finding an optimum compromise between the choice of scientific and technical terms, and the accessibility of the content to a wide readership: producers, trainers and agricultural extension workers.

This manual focuses on sustainable farming techniques to meet environmental challenges.

Finally, this manual is pragmatic. It has therefore limited the necessary abundance of detail.

Raphaël NGADI LITADI

Table of contents

CHAPTER 1: GENERAL INFORMATION AND VARIETAL SELECTION

1.1. Why do we grow corn?

Corn kernels are used to feed humans (as fresh ears of corn or as meal) and livestock. The starch extracted industrially from the kernels is used to make baby food, cookies, beer, glues, textiles, fabric finishes, etc.

Corn germs yield oil, which is used for human consumption and in the manufacture of margarines, soaps, varnishes, artificial textiles and more. It is also grown as green fodder for animals or to make silage for cattle.

1.2. Description

aize is a tropical plant of the grass family native to Central America (Mexico).

Photo 1: Flowering ear

1.3. Selection

Varieties are classified according to the length of their vegetative cycle on the one hand, and the characteristics of their kernels on the other (color, shape, etc.).

Thanks to crossbreeding facilities, breeders have obtained highly productive hybrids.

The creation of hybrid maize with a fixed formula has the

disadvantage of requiring the renewal of seeds for each crop, which necessitates a complex seed production organization. Maize breeding must take into account the difficulties involved in creating such an organization.

The selection formulas are therefore multiple and adapted to seed production conditions: (i) improved local varieties; (ii) introduced varieties; (iii) composites of local and/or introduced varieties; (iv) synthetic varieties resulting from the crossing of stabilized lines, and able to be exploited in advanced generation; (v) simple or complex hybrids.

The main selection criteria for corn are :

- Grain production and quality;
- Adaptation of the cycle (in regions with two rainy seasons, such as Gabon, we look for varieties that are 90 to 100 days early from sowing to maturity, allowing two crops to be grown per year);
- Plant architecture (the aim is to reduce maize's vegetative development);
- Resistance to lodging (the aim is to reduce the size of the corn and in particular to lower the position of the ear on the

stalk);

- Ear protection (we look for spathes that completely cover the ear to protect it);
- Tolerance to parasites.

Photo 2. View of field corn

Photo 3: View of hut corn

What you need to know

In humid equatorial environments such as Gabon, crops in general have a particular way of working, linked to very specific climatic aspects.

The almost permanent cloud cover means that global radiation levels are low all year round.

Day and night temperatures are uniform and do not change from one season to the next. At night, plants burn off much of the energy they have accumulated during the day, as a result of maximum nocturnal respiration.

The absence of an off-season adds to these natural conditions, favoring intense biological activity for crops and their enemies. This environment maintains a high level of general aggressiveness on the part of weeds, insect pests and diseases.

In this type of environment, there are no restrictions on the choice of varieties. You have to juggle with a crop calendar and harvest dates that depend on the plant's physiology and the farm's technical and economic imperatives.

CHAPTER 2: SOIL PREPARATION

Soil preparation can be done mechanically or manually. Maize is a demanding plant, very sensitive to variations in soil fertility, and responds well to fertilizer inputs, especially nitrogen.

It is particularly fond of soils rich in organic matter and with good physical properties.

2.1. Soil composition

Soil is composed mainly of clay, sand, limestone and humus. These different elements are always mixed in very different proportions, resulting in soils of varying degrees of fertility.

Clay soils are impermeable and retain moisture. They are ideal for corn, a water-demanding species (but avoid overly clayey soils). But clay soil is hard to work. A thick crust forms on its surface in dry weather.

Sandy soils are permeable and easy to work, but they dry out very quickly. They are suitable for drought-resistant plants such as peanuts. But they are usually poor soils.

Humus is very rich in plant nutrients. Compost and potting soil are humus obtained by leaving plant debris to rot for a certain length of time.

Humus makes sandy soil, which is naturally light and permeable, more compact. On the other hand, it aerates clay soil that is naturally too heavy, making it easier to work.

2.2. Mechanical preparation

Most dry-season ploughing is carried out with share ploughs. Given the soil's susceptibility to erosion in humid equatorial environments, maize reseeding is carried out exclusively with tine tools, either in a single pass with a vibrocultivator, or by interposing a pass with a chisel when the clods are too large. Ploughing with ploughshares or discs on very wet ground has a destructuring effect on the soil.

The harshness of the natural environment and the rapid degradation of cultivated plots mean that environmentally-friendly systems must be sought out and favored.

Photo 4: View of a Decompactor

Photo 5: View of a Chisel

2.3. Manual preparation

The land used may be cleared bush or forest. If necessary, ploughing will be carried out using a hoe after the plot has been cleared.

Amendments and fertilizers should be applied to the seed rows before ploughing. But it's important to assess the need for them in the light of soil analysis results, and the benefits to be gained from their use.

In a first crop, in a forest clearing or in old fallow land, these inputs can be useless and unprofitable. In fact, the soil may be sufficiently rich to enable higher yields to be obtained.

Photo 6: View of a plot of land under manual preparation for garden corn (ploughing)

Photo 7: View of a field being prepared manually for field corn

2.4. Amendment

Amending soil means improving its physical properties.

Some soils are too hard (clay), while others are too light (sand). These are corrected by soil amendments.

Good soil should contain 60-70% sand, 25-30% clay, 5-8% limestone and 4-5% humus.

Liming is the addition of pulverized lime (shells, eggshells, etc.) to correct excess sand or clay.

Gabon's soils are generally very acidic. This translates into a low pH, which few cultivated plants can withstand.

In order to achieve suitable agricultural production, we need to find a way of mitigating this acidity. Magnesium lime and dolomite are commonly used.

These amendments provide the calcium and magnesium needed to "raise the pH".

2.5. Fertilization

Fertilizing soil means transforming its composition to provide what it lacks. Compost or manure adds the nutrients required by plants to any soil.

2.5.1. Fertility thresholds

The fertilization thresholds given below are based on the results of nitrogen fertilization trials carried out in the 80s on SIAEB farms in Gabon.

Nitrogen is a key element in plant growth. For corn, 110 kilograms of urea per hectare should be applied twice, three weeks after sowing and at flowering.

This practice is satisfactory in good growing and climatic

conditions. However, any delay in the application date due to rainfall results in a drop in yield.

2.5.1.2. Magnesium

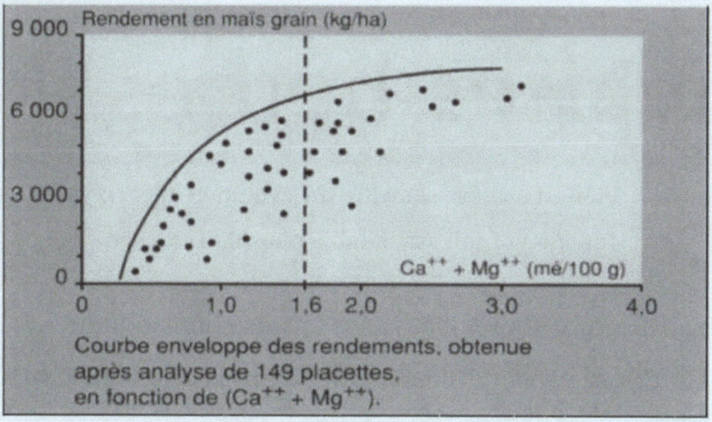

Figure 1: View of a maize yield curve as a function of soil calcium and magnesium levels.

N.B: This graph shows that if soil (ca + Mg) is less than 1.6 me/100g, yields are limited.

2.5.1.3. Phosphorus

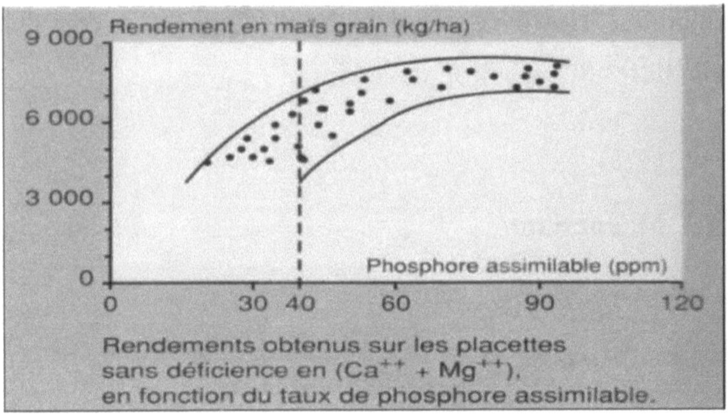

Figure 2: View of a curve showing the evolution of corn yields as a function of soil assimilable phosphorus levels.

N.B.: The graph shows that yields are also limited if the assimilable phosphorus content of the soil is below 40ppm.

2.5.1.4. The balance between potassium and magnesium

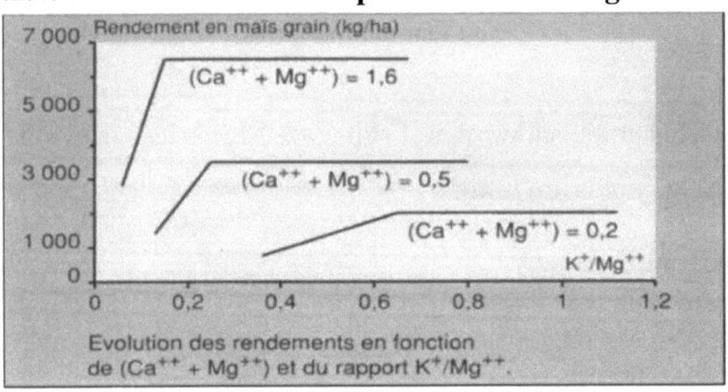

Figure 3: View of the evolution of maize yields as a function of the potassium/magnesium ratio.

N.B.: The effect of the ratio K^+/Mg^{++} is all the more marked the higher the yield potential, in relation to the medium of

$(Ca^{++} + Mg).^{++}$

2.5.2. Organic fertilization

On the whole, mechanized and even manual operations enable all crop residues to be buried. But when a plot of land seems to be "out of breath", spreading 20 to 30 tonnes per hectare of fresh chicken droppings gives a spectacular boost to yields.

What you need to know

Gabon's soils are generally very acidic. This translates into a low pH, often between 5 and 5.5, which few cultivated plants can withstand.

If we are to produce a decent crop, we need to find a way of reducing this acidity.

Spreading magnesium lime and dolomite is the main solution.

In some cases, where the pH is relatively high, using dolomite alone may be sufficient, as it already contains lime.

These amendments provide the calcium and magnesium needed to "raise the pH".

Other obstacles also arise, such as the lack of phosphorus and the balance between the levels of mineral elements required for crop nutrition.
So we need to define the types and doses of fertilizers that will correct these defects.

But the limit of fertilization to be used depends on the purchase price of fertilizers compared with the selling price of the extra grain obtained.

CHAPTER 3: CULTURE

In Gabon, the cultivation methods most often encountered are hut and open field. Irrigated farming is still the exception.

3.1 Cultivation techniques

Maize is very sensitive to improvements in soil physical properties. Yield increases due to ploughing alone, which in particular increases soil water reserves, are generally high and commonly reach 2%.

Depending on the region, the plant is grown flat, as in Gabon, on ridges or mounds. Flat, every 80 cm, a 10 cm-deep hole is made with a stick, into which 3 to 4 seeds are placed and covered with 2 cm of soil.

It is recommended to sow early, at the start of the rains. Any delay will result in lower yields.

Optimum stocking density is generally between 40 and 60,000 plants per hectare. We recommend using
spacing of 0.80x25, i.e. a density of 50,000 plants/ha. This represents a seed requirement of around 25 kg.

The optimum density value depends on :

- The variety used (higher density with low vegetative growth varieties);
- Water supply conditions;
- The level of soil fertility.

Photo 8: View of direct seeding of maize

What you need to know

Prior to clearing, the soil has a lumpy, runoff-resistant structure, ideal for extensive root colonization.

When cultivated, the cloddy state disappears and the soil is totally destructured. It becomes massive and highly susceptible to erosion.

To rebuild a new physical equilibrium, the use of certain ploughshares, and especially tines, is essential. Direct seeding techniques also preserve the soil's characteristics.

Heavy mechanization is to be avoided, as it induces enormous changes in the soil's physical properties.

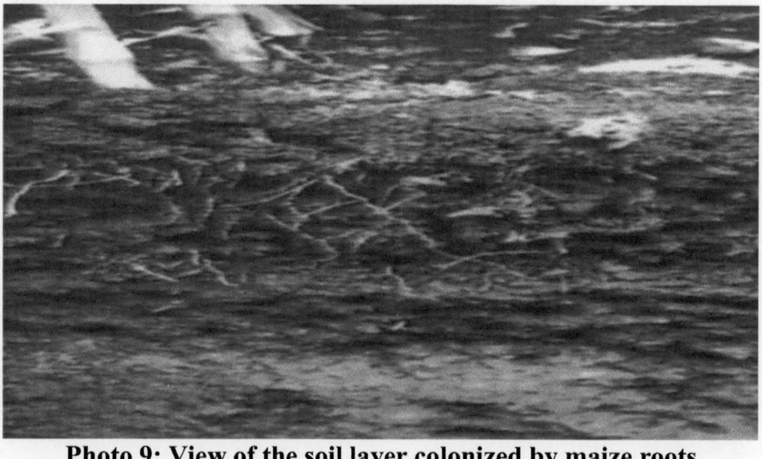

Photo 9: View of the soil layer colonized by maize roots

3.2. Maintenance

Weed invasion penalizes production. Corn yields drop sharply as weed cover increases.

Herbicides are used, but the results are not always certain.

As for cultivation methods, ploughing with share ploughs reduces the importance of weeds compared to tine ploughing.

However, the latter provide better yields when chemical products are effective.

Seedlings can be covered with ash as soon as they emerge from the ground, to keep insects away.

For manual maintenance, weed regularly, then, after a month, remove the weeds, leaving only two stems.

Photo 10: Field corn in full growth

3.3. Yields

Yields vary according to cultivation method. Traditional cultivation yields between 500 and 800 kg/ha, while field cultivation yields between 3 and 4 tonnes per hectare.

In very good station conditions, 6 to 7t/ha are achieved.

3.4. Harvest

Corn is never harvested dry: its presence on the cob, which is always wetter than the grain, prevents its humidity from dropping to 12%, the level required for good preservation.

The grain must therefore be dried. Either the ears or the kernels can be harvested.

Ear harvesting is mainly manual in Gabon (25 working days/ha), but can be mechanized.

Grain harvesting is carried out mechanically using an appropriate machine.

Photo 11: View of a pile of corn cobs after harvesting

What you need to know

Maize can be produced in a crop rotation system: maize/soya or maize/peanut, for example. But also, in a system integrating a fallow: corn/fallow. These successions are coupled with ploughing to work the soil.

Corn/Soybeans: soybeans are grown after a corn cycle, direct-seeded into corn debris to avoid destructuring the soil with successive ploughing. Ploughing is carried out after the soybean cycle.

With short-cycle varieties, it is possible to grow two crop cycles per year.

Maize/fallow: this treatment involves leaving plots fallow behind maize. In this case, the aim is to use longer-cycle varieties.

We must remain vigilant when it comes to maintaining fallow land, and intervene before weeds seed.

Dry-season ploughing can be carried out quite early, firstly because the plot is not under cultivation, and secondly because the fallow dries out the cultivation profile very vigorously.

Vocabulary

Priming: Operation performed on raw materials before they are processed or presented.

Weed: Plant added to a stand of vegetation without being deliberately sown, to which it is initially foreign and whose presence is undesirable.

Amendments : Substances incorporated into the soil to improve its physical properties.

Nitrogen, phosphorus, potassium and magnesium: Macro-elements essential for plant development, germination and seed maturation.

Chisel: agricultural equipment used to scarify the soil.

Plough: An agricultural implement used to plow fields.

Dethatching: Operation which consists of manually removing some of the young plants.

Erosion: Wear and transformation of soil by water.

Grumele: A granular material.

Hybrid: Corn produced by crossing different varieties or species.

Lineage: Set of individuals descending from one or more parents. In plants, the lineage is the result of successive self-fertilization over several generations.

Parasite: an organism that lives at the expense of another living host organism, on or in which it finds sustenance and protection.

Ph: Abbreviation for hydrogen potential. A measure of the acidity or basicity of a solution.

Runoff: The flow of water over the earth's surface, especially the surface of soils.

Cob: The corn cob is the rachis of the corn cob, the female corn inflorescence on which the female spikelets are implanted, which turn into kernels as the plant matures.

Variety: Term used to designate the set of individuals within a given species, distinguished from any other set of plants of the same species by the expression of minimal hereditary differences determined by a given genotype. These particular characteristics can be reproduced from generation to generation by seeds.

Vibrocultivator: Agricultural tool for shallow soil cultivation, with vibrating tines, generally powered by a tractor.

Bibliography

Alain CAUJOLLE, 1992. Essai systèmes de culture intégralement mécanisés maïs/soja en milieu équatorial humide. Center de Recherches Agronomiques de Boumango (Gabon), 49p.

J. Ernoult, 1991. Agriculture et petit élevage en zone Tropicale. France: éd. Saint-Paul, 80p.

E. Torquebian. 2007. Agroforestry: trees and fields. Pp.26...

R. NGADI LITADI, 2010. Impacts des Activités Agricoles au nord de la zone tampon de la Réserve de Biosphère du Niokolo Koba. Master's thesis: Gestion durable des Ressources Naturelles: Université de Thiès (ENSA) (Senegal), 37p.

Ministère de la Coopération et du Développement 1991. Mémento de l'Agronome (fourth edition), pp 651-664.

G. Delafond et al. 1990. Synthèse des travaux effectués à la SIAEB (Gabon) par la cellule de Recherche d'Accompagnement pendant la phase II, IRAT/CIRAD, 28p.

Ange A., Casta P., 1990. Cellule de Recherche d'Accompagnement, périmètre SIAEB (Gabon), volet Agronomie. Rapport de synthèse des campagnes 1988-1990, Montpellier, France, RAT, 58p.

Herblot, 1986. Quelques problèmes posés par la motorisation intégrale des cultures annuelles (maïs et soja) en zone tropicale humide: Cas de la SIAEB au Gabon. Machinisme Agricole Tropicale, 95p.

Abderrazak MAROUF, Joël RAYNAUD, 2007. Botany from A to Z. Dunod, Paris, 2007. Imprimerie SNEL Grafics sa, Z.I. des Hauts-Sarts-Zone 3, Rue Fond des Fourches 21-B-4041 Vottem (Herstal), Belgium, 342p.

CAUJOLLE GAZET A., 1992. Tillage and cropping profile in mechanized cropping systems with annual maize/soybean succession in humid equatorial environments. Le cas de la SIAEB au Gabon. Boumango, Gabon, CRAB, 25p.

APPENDIX

As an indication, a hectare of corn can cost 440626 FCFA (671 euros).

With a yield of 3 tonnes per hectare and a field price of 200 FCFA per kg, the profit is 159374 FCFA (243 euros).

Designation	Expenses			Products		
	Quantity	Price	Amount	Quantity	Price	Amount
Production sales				3000	200	600 000
Seed purchase (kg)	25	500	12500			
NPK	150	500	75000			
Urea	100	500	50000			
Organic manure	5	13000	65000			
Ploughing and spraying	1	20000	20000			
Lubricant and fuel			57500			
Workforce (HJ/month)	25	5000	125000			
Harvest			20000			

Depreciation and amortization			**15626**			
Total			**440626**			**600 000**
Net operating income			**159374**			

Example of an operating account for one hectare of corn

Biography:

Raphaël NGADI LITADI

Gabonese national, Master's degree in Forestry and Environment for Sustainable Management of Natural Resources, Diploma in Agricultural Engineering, Baccalauréat série D.

I worked as a civil servant on secondment for 11 years at the Société de Développement de l'Hévéa Culture where I held the positions of Head of the Agricultural Department and National Director of the Village Program.

Back at the Ministry of Agriculture, I successively held the positions of Head of the Agricultural Research Department, Deputy Director General of Agriculture, Acting Director General of Agriculture, Inspector General of Services, and currently hold the position of Technical Advisor to the Minister of Agriculture.

I take part in international discussions as Soil Focal Point of the Global Soil Partnership (GSP), as my country's representative in the dialogue on forests, agriculture and trade in raw materials (FACT) and as national coordinator for the transformation of agricultural systems (FAO).

Raphaël NGADI LITADI

Mail: ngadilitadi@yahoo.fr

MIX
Papier aus verantwortungsvollen Quellen
Paper from responsible sources
FSC® C105338

Printed by Books on Demand GmbH, Norderstedt / Germany